THE KING'S NOSE

An Ecological Fairy Tale

by

Jeannie Morrison Walton

Illustrated by

Jeane Wallace Whittenburg

AuthorHouse™
1663 Liberty Drive
Bloomington, IN 47403
www.authorhouse.com
Phone: 1-800-839-8640

First published by AuthorHouse 01/01/2012

ISBN: 978-1-4678-7067-2 (sc)

Library of Congress Control Number: 2011961780

Printed in the United States of America

This book is printed on acid-free paper.

To my grandson, Alexander,
who renamed himself, Xan
Love from Nana

Special thanks to my Massachusetts cousin, Joe
Scott, who pointed me in the right direction and
gave encouragement.

To Rachel Carson, 1907-1964, author of *Silent
Spring,* who lit the candle and led the way to
ecologically preserving our balanced Earth.

To Lady Bird Johnson, 1912-2007, for her lifetime
achievements in beautification and conservation.

Once in a beautiful little kingdom lived a king. Through the kingdom flowed a quiet river. Beside the river grew an orchard. The orchard was full of fruit trees where the people of the kingdom harvested a bountiful crop of apples, peaches, pears and plums every year.

In the spring the trees were decked with an abundance of fragrant flowers, and the ground under the trees was sprinkled with pollen. Bees and butterflies flitted through the garden pollinating the blossoms. The king ruled over a kingdom where the insects merrily did their job of pollinating all the plants so the harvests would increase their yield.

One day the little king was strolling through his royal rose garden, and he stopped to inspect a most magnificent bloom. The fragrance of the bloom filled the air with sweetness, and the king knelt down to take a big whiff of its extraordinary odor. His eyes were filled with delight as he inhaled its fragrance. Suddenly a startled bee buzzed out of the bloom and angrily stung the king on the tip of his bulbous nose.

"Guards!" shouted the offended king. "Guards! I have been attacked! Swat this horrible buzzing bee!" The king pointed to the ground where the bee lay dying.

"But, Majesty!" exclaimed Alexander, who was only twelve-years-old and the youngest-man-of-arms. "After a bee stings someone, he is not able to pull the stinger back out. The bee is already dying!"

Twenty men-at-arms rushed to the king, and the largest of the guards mightily swatted the bee.

The king's already large nose swelled to an enormous size. Tears streamed from his eyes, and he began to cry.

"Kill them! Kill them, every single bee!" sobbed the revengeful king.

"But, Majesty", cried the smallest man-at-arms. "If we kill all the bees, then what will pollinate the orchards and the gardens?"

"Pollinate! Smolinate!" shouted the king. "Who cares about such an absurd word? My poor nose has been attacked!"

Immediately, all the other men-at-arms began running through the royal rose garden swatting all the bees. By sundown not a single bee could be found in the royal rose garden.

The next morning the king took an enjoyable stroll through his royal rose garden and did not encounter a single bee.

Feeling very smug and proud of himself, he decided to spend the afternoon having tea and cookies in his royal orchard down by the sparkling river. After finishing his mid-day snack, the little king took a stroll through the orchard's spring blossoms.

Lost in the beauty of the blooms, the king did not hear the soft buzz of the bees as they flitted from flower to flower. He pulled down a flowering branch and started to take a sniff when he spied a bee supping on some nectar. In terror he let go of the branch and it snapped back into the air. The startled bee came after the closest object, which was the king. The king waved his royal hands and forbade the bee to sting, but the angry bee aimed for the largest and most unprotected part of the king, and it stung him soundly on his large red nose.

The little king began to run towards his men-at-arms all the while shouting and crying that he was again under attack. Alexander, the youngest man-at-arms, once more tried to explain to all the other men-of-arms of importance of bees in the pollination of blossoms. No one listened. The men-at-arms were too busy swatting and squashing all the bees in sight.

The next day the king issued a royal decree that every person in the kingdom was to spend the next two days swatting and killing bees.

The following week the king noticed a bee in his royal garden. Enraged, the king sent messengers all over the kingdom that a reward of twenty gold pieces would be given to whoever could wipe out the enemy of his royal nose.

Three days later a strange old man appeared at the door of the castle. The old man was tall and thin. He was dressed head to toe in dusty black with his robe cinched at his waist with a frayed rope. His face had an evil leer from a deep jagged scar on the left side of his cheek. His gray eyes glared from beneath dark shaggy eyebrows.

"Majesty," he whispered leaning close to the king. "I have the solution to your problem. But you must be warned! It will kill all things that fly through the air on wings"

"Yes! Yes!" shouted the excited king. "The very thing we need."

"No!" exclaimed Alexander. "Haven't you been listening to me? We need the bees and the butterflies to pollinate our gardens and orchards."

"Yes, yes, butterflies are very beautiful the way they flit around the flowers. Can we spare the butterflies?" the king questioned anxiously.

"Hear me!" hissed the old wizard, for wizard he was. "I have a special spray that will float through the air and kill all things that fly."

The little king thought for a few moments, then turned angrily to the littlest man-at-arms who was tugging at his royal robes and pleading with the king not to trust this wicked wizard.

"Enough!" the king shouted at Alexander. "I will rule in favor of my hurt royal nose. I will gladly pay to be rid of those bees and flying creatures forever!"

The next morning all the people of the kingdom hid in their houses and locked their doors because the evil wizard was going to begin his magic.

Alexander, the littlest man-at-arms, peeked through a crack in his door and saw a purple mist begin to fog the kingdom. Its airy tentacles drifted across the royal rose garden, down the hillside past the villagers' gardens, and on towards the orchards. Alexander began to cough as the odor of the mist crept into his room. He ran and hid his head under a pillow. Alexander fell asleep and dreamed of the death of all the flying animals.

The following morning was strange. Nothing was flying in the still morning air. Not a bee, not a butterfly, nor a bird could be seen.

"No!" cried Alexander. "What will happen to our beautiful kingdom without the insects and birds to pollinate our plants?"

After the day of the evil purple mist, the flowers on the orchard trees and the flowers of the gardens bloomed, wilted, withered, and fell to the ground without producing their life giving fruit. Without the fruit, no seeds for new plants were formed. The circle of life was broken.

That autumn there was very little harvest. Few apples, peaches, pears, or plums were picked from the orchard. The gardens had few beans, pumpkin, or squash. Wheat was in short supply. The only pollination was from the wind. It was a hungry winter, and the people began to think about what Alexander had tried to tell the king about the value of insects and birds to pollinate the blossoms. The children began to dream of the sweet honey that the bees had made. Even the little king began to wish for the delicious honey to spread on his breakfast pancake.

Together the people of the little kingdom marched to the king chanting, "Bring back the bees and birds! Bring back the birds and bees!"

The king looked out at the hungry faces of his people and realized the wrong he had committed against nature. Ashamed to look at the angry faces the little king asked, "How can I right this wrong? What can we do to bring our orchards, gardens, and my royal rose garden back to fruitfulness?"

The people began to talk among themselves and discus how best to bring back the balance of nature.

Alexander, the littlest man-at-arms, climbed upon a large rock so he could be heard above the angry people.

"Listen to me!" Alexander shouted.

All the people became quiet.

"I will cross the rivers, climb the mountains, and search until I find the bees and the other insects of the air. I will carefully capture them and return home with them."

An elderly lady asked, "But, what about the birds?"

"Grandmother, where there are juicy insects the birds will follow." replied Alexander.

So the littlest man-at-arms gathered twenty glass jars, punched holes in their lids, and tied them around his waist. He waved good-by to the little king and all the people and began his quest.

Alexander crossed a river and climbed a mountain. He crossed a watery swamp. Then he climbed a small hill until he came to a pine forest. There, to his delight, he saw a swarm of bees drinking nectar from an enormous bush covered with yellow bell shaped flowers. Alexander could smell the sweet fragrance of the flowers from where he stood. The bush was growing in a sunny clearing surrounded by native wildflowers of purple and gold. Dozens of brightly colored butterflies danced over the flowers.

Immediately, Alexander began to fill his jars with hundreds of the fast moving insects. Carefully, he scooped up the queen bee to rule over the new hive.

Soon all the jars, but one, were filled with buzzing bees or flitting butterflies. Alexander gave a mischievous little laugh and put four houseflies, two gnats, three dragonflies and five hungry mosquitoes into the last jar.

When he returned to the little kingdom the people and the king all shouted with joy as Alexander opened the jars and the insects came tumbling out.

He saved his special jar for last, and when he opened it the four houseflies, two gnats, three dragonflies and five hungry mosquitoes flew out. The biggest of the mosquitoes bit a plump farmer on his right cheek. The people all laughed as he tried to slap it away.

The next morning the king woke up and went for a walk in his royal rose garden. He stopped to smell an enormous red rose. As he leaned over to sniff, out flew a startled bee and stung him on his royal nose.

"Ah," sighed the little king as he rubbed his swollen nose. "The joys of nature!"

Pollination

Pollination is the transfer of pollen and the beginning of the growth of a new seed. Without pollination plants can not reproduce a new plant.

Flowers have different parts. They have the male parts which are called stamens. These stamens are topped with sticky pollen that has to be transferred to the female part called a pistil. The top of the pistol is called the stigma. Many times the stigma is sticky to help catch the pollen. The base of the pistil is called the ovule. The seeds are formed in the ovule.

Self-pollination is when a flower pollinates itself.

Cross pollination is when flowers are pollinated by animals or wind. The insect, bird, or bat drinks the sweet nectar and inadvertently gets pollen on its body. When it moves to the next flower it deposits the pollen on the new pistil. This type of pollination produces stronger seeds. However, apples only pollinate apples and peaches only pollinate peaches. Pollen from a rose can not pollinate a pear.

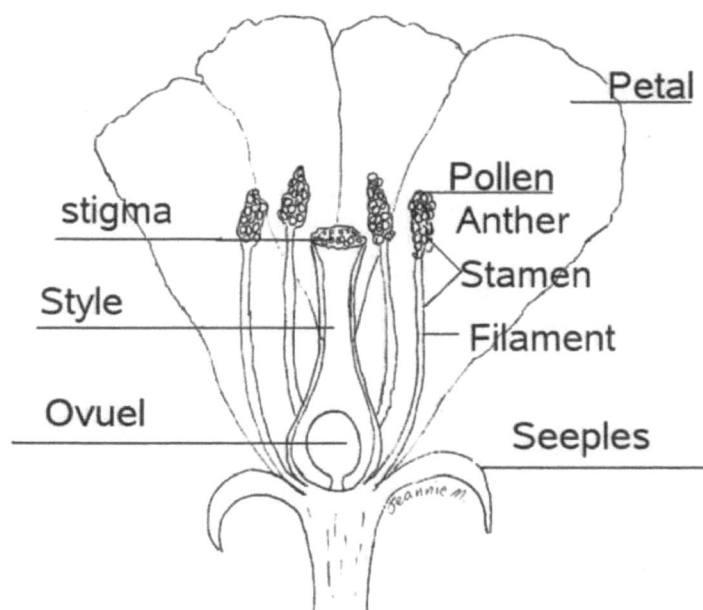

Many farmers put beehives in their orchards to pollinate their trees and other crops. One-third of the food we eat is dependent on pollinators.

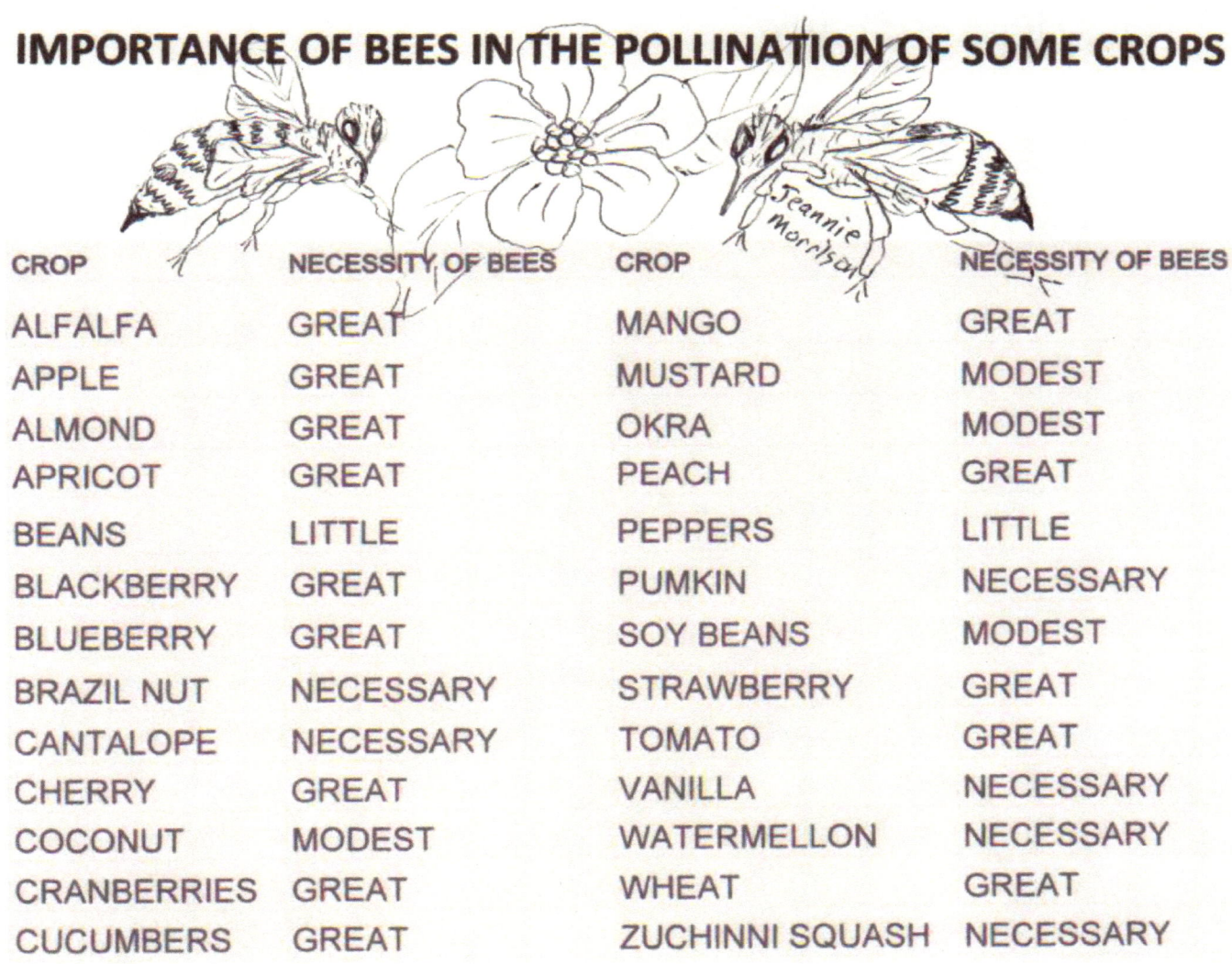

IMPORTANCE OF BEES IN THE POLLINATION OF SOME CROPS

CROP	NECESSITY OF BEES	CROP	NECESSITY OF BEES
ALFALFA	GREAT	MANGO	GREAT
APPLE	GREAT	MUSTARD	MODEST
ALMOND	GREAT	OKRA	MODEST
APRICOT	GREAT	PEACH	GREAT
BEANS	LITTLE	PEPPERS	LITTLE
BLACKBERRY	GREAT	PUMKIN	NECESSARY
BLUEBERRY	GREAT	SOY BEANS	MODEST
BRAZIL NUT	NECESSARY	STRAWBERRY	GREAT
CANTALOPE	NECESSARY	TOMATO	GREAT
CHERRY	GREAT	VANILLA	NECESSARY
COCONUT	MODEST	WATERMELLON	NECESSARY
CRANBERRIES	GREAT	WHEAT	GREAT
CUCUMBERS	GREAT	ZUCHINNI SQUASH	NECESSARY

There are many other seeds that when planted grow into a plant that will produce food for one year, but need pollination to produce seed to plant for the following year, Here are a few:

Beet	Broccoli	Brussel sprouts	Cabbage
Carrots	Celery	Onion	Turnip

The End

www.ingramcontent.com/pod-product-compliance
Lightning Source LLC
Chambersburg PA
CBHW051110180526
45172CB00002B/851